discovermore

A Look at Weather and Climate

Weather and Climate

Marie Harts

Britannica Educational Publishing

IN ASSOCIATION WITH

THE ROSEN PUBLISHING GROUP

Published in 2025 by Britannica Educational Publishing (a trademark of Encyclopædia Britannica, Inc.) in association with The Rosen Publishing Group, Inc.
2544 Clinton Street, Buffalo, NY 14224

Distributed exclusively by Rosen Publishing.
To see additional Britannica Educational Publishing titles, go to rosenpublishing.com.

Editor: Brianna Propis
Book Design: Michael Flynn

Photo Credits: Cover Piyaset/Shutterstock.com; (series background) Dai Yim/Shutterstock.com; p. 5 (top) salajean/Shutterstock.com; p. 5 (bottom) Jan Martin Will/Shutterstock.com; p. 6 Drew McArthur/Shutterstock.com; p. 7 Sakurra/Shutterstock.com; p. 8 Georgii Shipin/Shutterstock.com; p. 9 https://commons.wikimedia.org/wiki/File:Barometer_mercury_column_hg.jpg; p. 11 VectorMine/Shutterstock.com; p. 12 A3pfamily/Shutterstock.com; p. 13 Rashid Valitov/Shutterstock.com; p. 14 Designua/Shutterstock.com; p. 15 Leitenberger Photography/Shutterstock.com; p. 16 Eray Bozkurt/Shutterstock.com; p. 17 Vaclav Sebek/Shutterstock.com; p. 18 Designua/Shutterstock.com; p. 19 FaRifo/Shutterstock.com; p. 20 EAKARAT BUANOI/Shutterstock.com; p. 21 VectorMine/Shutterstock.com; p. 22 Edler von Rabenstein/Shutterstock.com; p. 23 Vixit/Shutterstock.com; p. 25 (top) Alexxxey/Shutterstock.com; p. 25 (bottom) NASA images/Shutterstock.com; p. 27 (top) SpiritArt/Shutterstock.com; p. 27 (bottom) Lorcel/Shutterstock.com; p. 28 Aleksandr Lutcenko/Shutterstock.com; p. 29 Maridav/Shutterstock.com.

Library of Congress Cataloging-in-Publication Data

Names: Harts, Marie.
Title: Weather and climate / Marie Harts.
Description: New York : Britannica Educational Publishing, in Association with Rosen Educational Services, 2025. | Series: Discover more! a look at weather and climate | Includes glossary and index.
Identifiers: ISBN 9781641901352 (library bound) | ISBN 9781641901345 (pbk) | ISBN 9781641901369 (ebook)
Subjects: LCSH: Weather--Juvenile literature. | Climatology--Juvenile literature.
Classification: LCC QC981.3 W347 2025 | DDC 551.5--dc23

Manufactured in the United States of America

CPSIA Compliance Information: Batch #CSBRIT25. For further information contact Rosen Publishing at 1-800-237-9932.

Contents

Weather vs. Climate 4
It All Starts with the Sun 6
Under Pressure 8
Water Vapor in the Sky 10
Defining Climate 14
Parallels 16
Water Vapor and Currents 18
Decreased Temps, Increased Heights 22
A Variety of Climates 24
The Role We Play 28
Glossary 30
For More Information 31
Index 32

Weather vs. Climate

Look outside your window. What is the weather like today? Is it sunny or cloudy? When you step outside, does the air feel hot or cold? Weather is the daily state of the atmosphere, or air, in any given place. Weather is made up of specific conditions. These include temperature, wind speed and direction, and precipitation. Weather conditions affect the comfort, food supply, and safety of people in all parts of the world.

Climate is not the same as weather. Climate is the average weather conditions in an area over a long period of time. Weather can change from day to day or even hour to hour. It also changes with the seasons. In the winter an area may be cold and snowy, but in the summer that same area can be hot and humid. Climate does not change so quickly. An area's climate is what the weather is like there most of the time, even when it changes with the seasons. Each season has about the same conditions every year, depending on an area's climate.

Water that falls to the ground from clouds is precipitation. Snow is one type of precipitation.

compare and contrast

How might your life look different during a day in winter versus a day in summer? What adjustments are made to match the weather?

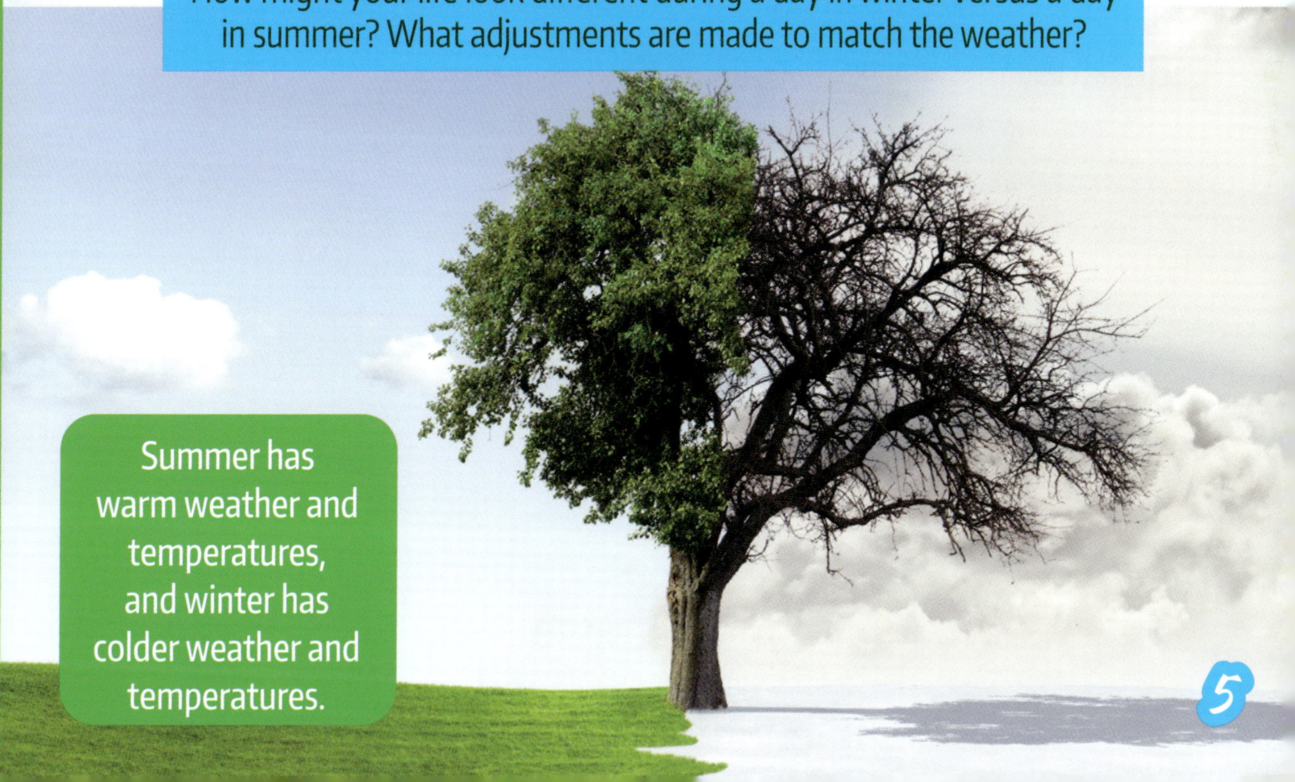

Summer has warm weather and temperatures, and winter has colder weather and temperatures.

It All Starts with the Sun

You may notice some elements of weather very easily. You can feel the temperature, which is how hot or cold it is outside. You can also feel the wind blowing. You know when it is raining or snowing too. Other weather conditions may be less noticeable. These include humidity and air pressure. Whether you can feel them or not, all of these conditions together make up the weather.

Wind is caused by an uneven heating of Earth's surface by the sun. Tropical zones are heated more than polar regions.

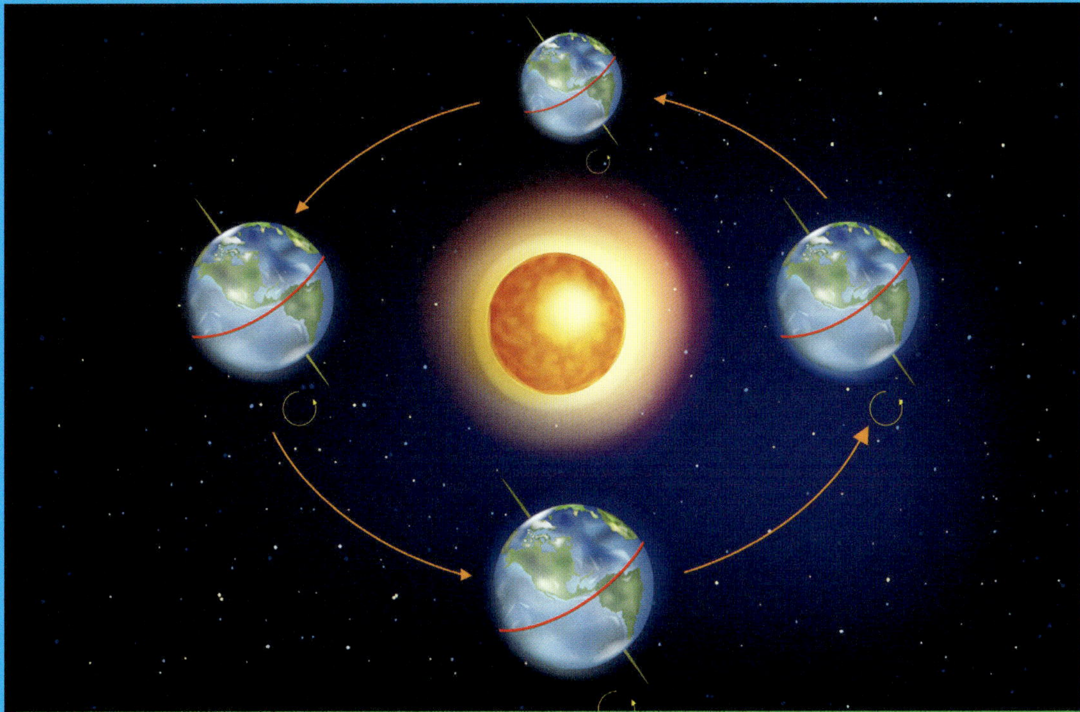

Earth's axis, or the invisible line that the planet rotates around, always points in the same direction. Due to this, the part of Earth that receives direct rays from the sun changes as Earth revolves around the sun.

The elements of weather all start with energy from the sun. The sun heats the air and the water. But the energy from the sun does not reach all parts of Earth equally or at the same time. This is why parts of the world experience weather differently from others.

Consider This

How does Earth's axis impact the seasons?

Under Pressure

An area's temperature is shaped by many elements. One is the amount of direct sunlight a place gets. Another is the type of landforms that are found in a particular place. Some parts of Earth's surface, such as bodies of water, absorb more energy than others. Temperature is also affected by how cloudy a place is. Clouds block some of the sun's rays during the day. But they can also help keep warm air near the ground at night.

Changes in air pressure help to predict local weather. Fair weather typically follows an increase in air pressure, and clouds and precipitation follow a decrease in air pressure.

Differences in heating affect Earth's atmospheric pressure. The air that surrounds Earth is called the atmosphere. Its weight presses down and creates atmospheric pressure. Heat from the sun causes changes in this pressure. That is because as air is heated it becomes lighter.

Areas of high and low pressure are always moving around the globe. Air in areas of high pressure always moves to areas of low pressure, causing a shift in the weather.

Atmospheric pressure is measured with a barometer. In this barometer, a column of mercury in a glass tube rises or falls as the weight of the atmosphere changes.

Consider This

If you fill a balloon with air and hold it closed, the air pressure inside the balloon is higher than outside of it. What happens once you let go of the balloon?

Water Vapor in the Sky

Wind is also a result of changes in temperature. For example, the sun heats land and water differently. Air over land is usually warmer than air over the water. As the heated air over the land expands, it becomes lighter and rises. Heavier air from the ocean blows in to take its place. This creates a sea breeze.

Winds are named for the direction they come from. For example, a westerly wind starts in the west and moves east.

Water also plays a part in the weather. Water in oceans, lakes, and rivers **evaporates** when it is heated by the sun. This water moves into the air as a gas called water vapor. Humidity is the amount of moisture (that is, the amount of water vapor stored) in the air. Warm air can hold more water vapor than cold air. This is why summertime air feels moister and more humid than in the winter, when the air feels dry.

WORD WISE

EVAPORATION IS THE PROCESS WATER UNDERGOES WHEN TURNING FROM LIQUID INTO VAPOR AFTER BEING HEATED.

During the day, the land's surface heats faster than the sea's surface because the breeze is coming from the sea. But at night, the sea's surface is warmer because the breeze is now coming from the land. The land's surface cools quicker than the sea's at night.

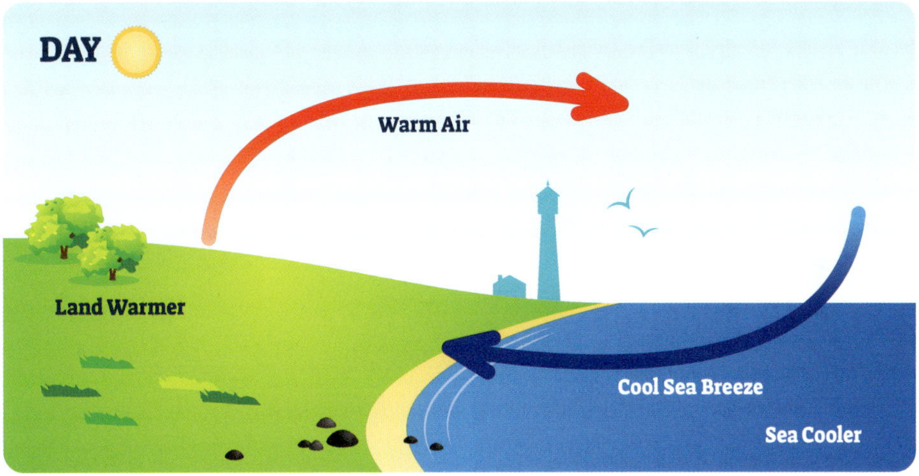

compareandcontrast

How is the air over an ocean different from the air over land?

Water vapor cools as it rises into the sky. That is because the temperature of the air is generally cooler as one moves away from the surface of Earth. The vapor starts to turn back into a liquid in the form of water droplets. The water droplets together form clouds. When many droplets come together, they form larger, heavier drops. At some point they get too heavy to stay in the cloud and they fall to Earth as precipitation.

Precipitation occurs when the air reaches its saturation point, meaning it's too full of moisture. When this happens, the moisture condenses on particles, such as dirt or dust, and forms raindrops.

Freezing rain is a special type of precipitation. It forms when rain falls and freezes as it hits an object.

There are many different types of precipitation, depending on the temperature of the air. If the temperature is warm, water droplets in the clouds will fall as rain. The droplets freeze and form snowflakes when it is cold.

Defining Climate

Weather conditions throughout the world are tracked and studied by **meteorologists**. They describe the climate of an area by comparing the measurements of certain conditions over a long time. For example, they have kept track of the temperatures for every day in the last 100 years in Chicago, Illinois. They are able to tell us that the temperatures in September are usually between 57° and 76° Fahrenheit (14° and 24° Celsius).

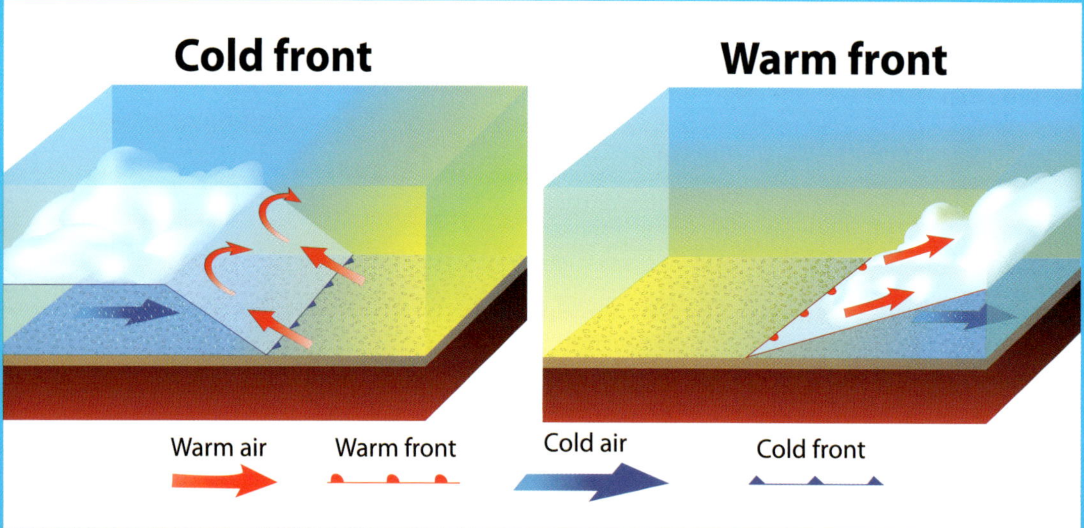

This illustration shows weather fronts. Cold fronts occur when cold air forces warmer air to rise. Warm fronts occur when warm air rises up and over cooler air.

Radar and satellites are two key technologies used by meteorologists. Radar is used to locate storms and help predict the weather. Satellites take pictures and send back information about the weather conditions happening all around Earth.

The weather conditions in an area can change a great deal from year to year, but over a long time patterns can be seen. The patterns show that weather conditions occurring in one part of one year are very similar to weather conditions that occur in other years during that same time. The highest and lowest temperatures ever recorded at a site are also important to know. They help people to know the range of temperatures to expect in a location.

WORD WISE

A METEOROLOGIST IS A SCIENTIST WHO STUDIES THE WEATHER. THEY USE COMPUTERS TO STUDY THE INFORMATION THEY COLLECT.

Parallels

For weather in an area to be the same each year, certain features have to be in place. One of those features is the latitude of a place. Lines of latitude (also called parallels) circle Earth parallel to the **equator**. The lines describe how far north or south of the equator a place is located. The latitudes closest to the equator are called the tropics. They are almost always warm because they get direct sunlight all year long. Places that are far away from the equator do not get a lot of direct sunlight. These include places like the North and South Poles, which have cold temperatures.

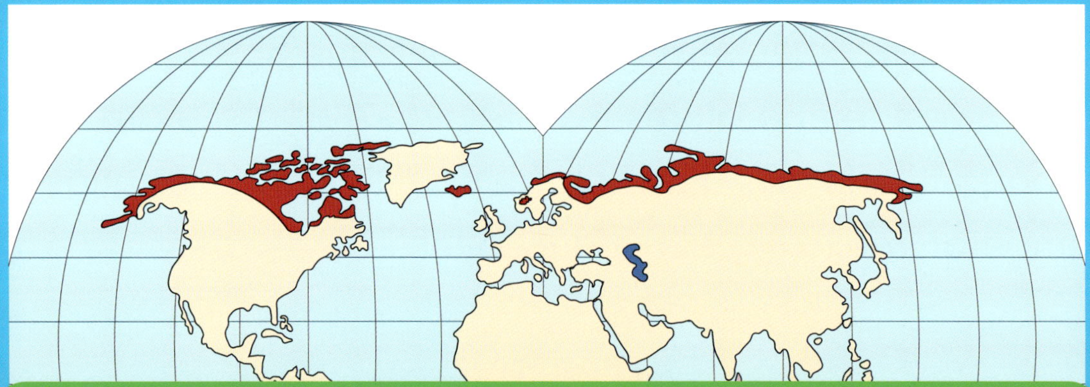

As latitude increases, the average temperature cools. This is why the tundra—which occurs between 60° and 75° of latitude—is very cold and dry.

The polar bear lives in the polar region—as you may have known from its name! The polar region has a latitude of 66°.

WORD WISE

THE EQUATOR IS AN IMAGINARY CIRCLE AROUND EARTH THAT DIVIDES THE PLANET IN TWO EQUAL HALVES. THE NORTH POLE IS IN THE TOP HALF, AND THE SOUTH POLE IS IN THE BOTTOM HALF.

Water Vapor and Currents

Oceans have a large impact on an area's climate. Most of the sun's heat is absorbed by the oceans. As the oceans warm up, water evaporates and turns into water vapor. The water vapor makes the air warm and wet. Because of this, land near an ocean usually has a milder climate than land away from the ocean.

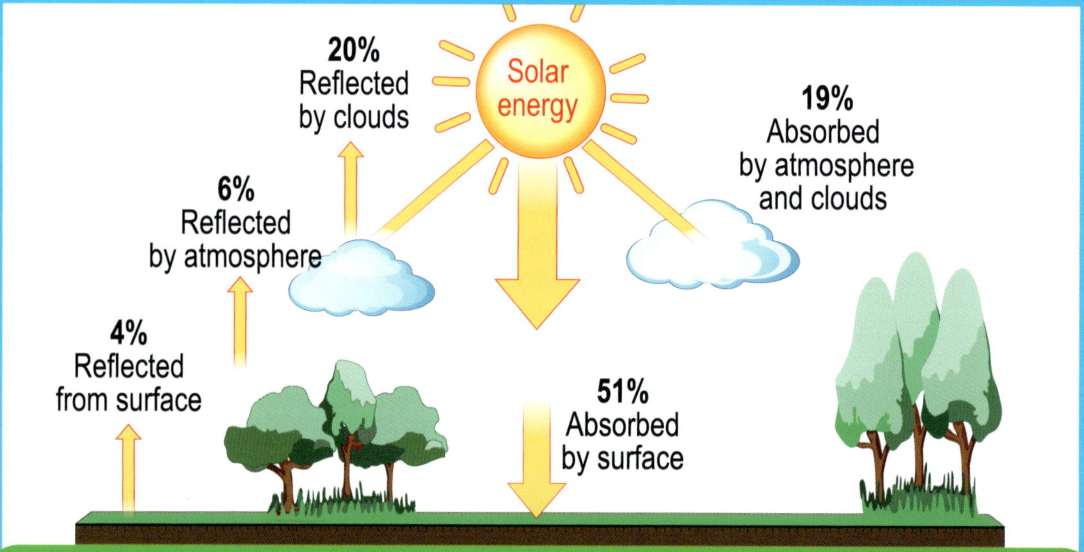

Seventy percent of the incoming solar radiation is absorbed by the atmosphere and Earth's surface, and 30 percent is reflected back to space.

Tropical ocean temperatures range from 72°F to 82°F (22°C to 28°C) depending on the time of year, and Arctic ocean temperatures range from 34°F to 37°F (1°C to 3°C).

Ocean currents also play an important role in climate. Winds move warm water from the tropics to the poles and cold water from the poles toward the tropics. Some of the major currents move in set patterns. For example, the Gulf Stream is a collection of currents that flow around the edges of the northern Atlantic Ocean. It brings warm waters to northern areas and colder waters to the south. Cities near the coasts will thus be warmer or colder than areas farther inland.

Consider This

The state of Alaska is much farther north than the state of Florida. The average summer temperatures in Anchorage, Alaska, are a lot lower than Tampa, Florida—67°F versus 90°F (19°C versus 32°C). Why might that be?

Certain wind patterns impact climate as well. Warm, moist tropical winds blowing from east to west are known as easterlies. They bring constant pleasant weather to the Hawaiian Islands and other tropical locations. These winds are also known as the trade winds. Long ago they helped power the sailing ships of traders. Winds that change direction twice a year are called monsoons. They often bring wet summers and dry winters to the regions where they blow. Monsoons most often occur in South Asian countries, such as Vietnam, Thailand, Cambodia, Bangladesh, Laos, India, and Pakistan.

Monsoons are caused by temperature differences in the air over land and sea. They bring heavy rain and often cause flooding.

This map displays global wind direction. Easterlies move from east to west, and westerlies move from west to east. Easterlies are warmer than westerlies because they're closer to the equator.

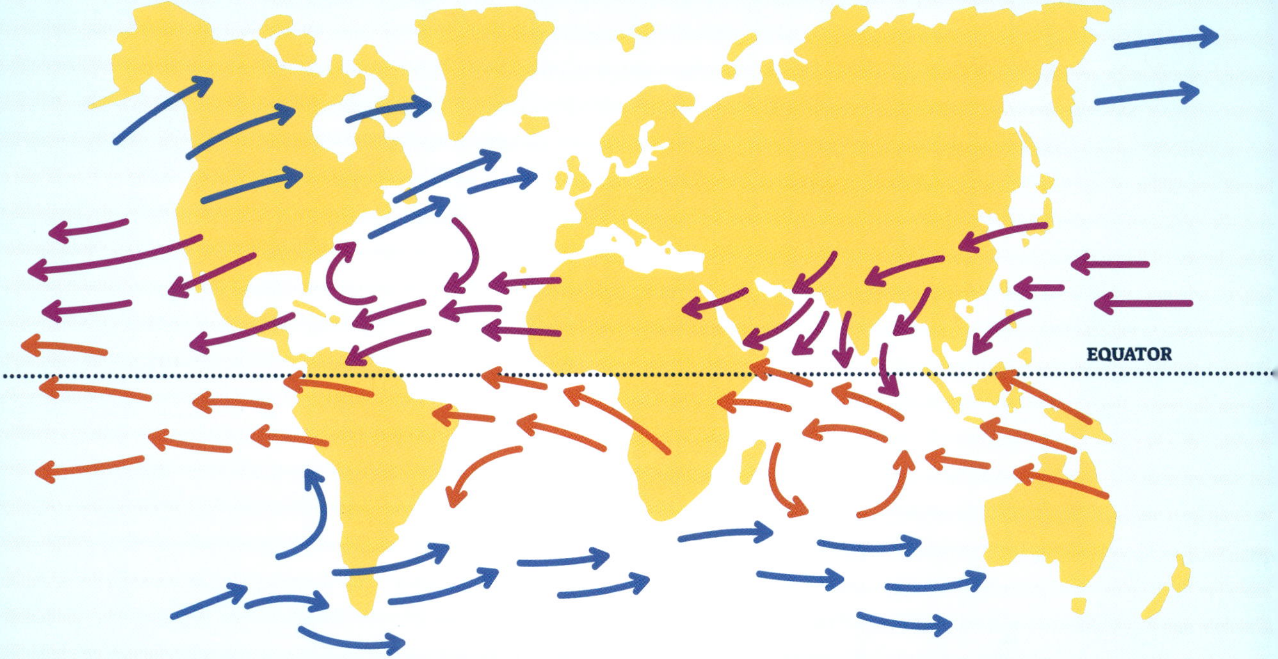

compareandcontrast

Winds can be categorized as either local or global. Which kind is a sea breeze? A monsoon?

Decreased Temps, Increased Heights

Since **altitude** affects temperatures, it also affects climate. Temperatures decrease as altitude increases. For example, land high up on a mountain is cooler than land closer to sea level. Even in tropical areas the weather can be cool for people who live in the mountains. High mountains near the equator, for example, may have tropical plants at their bases but ice and snow at their tops.

The snow on the Swiss Alps, seen here, demonstrates that the mountain peaks are colder than the land around them—even if the weather is not necessarily cold.

The point on Earth with the highest altitude is Mount Everest, located in the Himalayan Mountains on the border of Nepal and Tibet.

Mountains can also block cold air from traveling into an area. This creates a warmer climate. Mountains keep some areas drier than others as well. They can block rain and snow from falling. One side of a mountain may have dry land, while the other has lush land.

WORD WISE

ALTITUDE IS THE HEIGHT OF SOMETHING ABOVE SEA LEVEL.

A Variety of Climates

Every place on Earth has a unique climate, unlike any other. However, individual factors combine to form several general kinds of climates.

Tropical climates are warm all year. Some tropical climates have a lot of rain. Others are dry. Tropical climates are found near the equator.

Subtropical climates are found north and south of the tropical climates. These have a greater range of temperatures than tropical climates. They also may be humid or dry.

Cyclonic climates are found mostly north of the equator. In these climates, cold air from the poles mixes with warm tropical air from near the equator. This mixing often causes rain and snow, as well as warmer summers and colder winters.

Desert climates are warm and experience little to no precipitation.

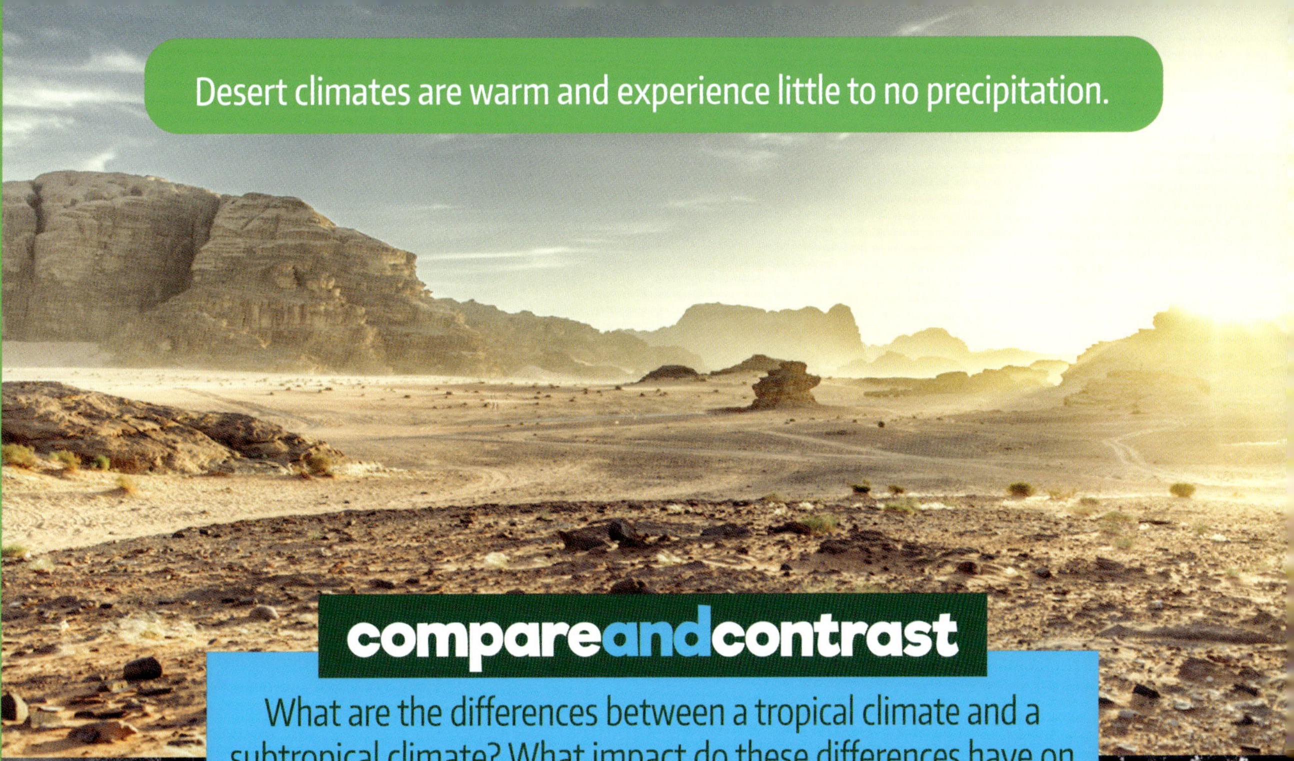

compareandcontrast

What are the differences between a tropical climate and a subtropical climate? What impact do these differences have on the weather in those areas?

How many cyclones can you spot in this satellite photo from space?

Snow and ice often cover polar areas. The polar climate is very cold. Some polar areas always have a layer of frost below the soil. This layer is called permafrost.

Highland climates are cooler than lowland climates on the same line of latitude. Highland climates have a great range of temperature between day and night. They tend to be more humid and cooler than the lower lands nearby.

Within each of the general climate types there is a great deal of variety. For example, the island of Hawaii covers about 4,000 square miles (10,360 sq km) in the Pacific Ocean. It is near the equator so it has a tropical climate in general. However, the eastern side of the island gets much more rain than the western side. If you climb one of Hawaii's mountains, you will find that the top of the mountain is cold and dry. Parts of the island near the ocean, however, are warm and wet.

Like desert climates, polar climates cannot support much plant life.

compare and contrast

If you lived closer to the equator, how would your daily life be different than if you lived farther from the equator? How might it be the same?

Since Hawaii is located in the tropics, the length of day and the temperature range don't change much throughout the year.

The Role We Play

Weather and climate affect each person on Earth in numerous ways. Weather affects what clothes people wear, how we spend our time, and how we travel.

Climate determines what kind of food can grow in an area and how much energy we use to stay warm or cool. Climate even determines where people can live. If a climate is too cold or too hot, too wet or too dry, certain animals and plants cannot survive there.

Mysterious holes began appearing in the ground in Siberia in 2014. Researchers believe they were caused by the thawing of permafrost—a result of increased temperatures—and methane gas trapped below the surface.

Most cars burn fuel that pollutes the air. Car emissions can even cause acid rain, which harms ecosystems and the organisms living there.

People can also have an effect on the climate. We burn fuels such as oil and coal to run our cars and power our homes. When we do that, the fuels release gases into the air. These gases trap heat on Earth. Scientists have discovered that this is making Earth warmer over time. This is called global warming. Global warming could lead to changes in our everyday weather and our climate. Storms may be more powerful, for example. Global warming can also cause polar ice caps to melt. This would cause sea levels to rise. Plants, animals, and buildings along coastlines would be in danger, and animals in Arctic regions would lose their homes.

Consider This

Burning fuel releases harmful gases into the air that cause air pollution and can contribute to global warming. What are some ways we can reduce the amount of these gases in the air?

Glossary

Arctic The regions around the North Pole with cold temperatures.

atmospheric pressure The weight of the air pressing down on Earth.

cyclonic Describes a type of climate that mixes cold and tropical air.

elements Components that describe the weather at any given time, such as temperature and wind.

emission The production and release of something, such as harmful substances.

energy The process of atoms and molecules moving faster and colliding with each other, generating heat.

front A border between areas of cold air and areas of warm air.

lush Having soil or vegetation that is rich.

moisture The presence of a liquid, such as water, in small amounts or condensed onto a surface.

noticeable Able to be seen.

radar A system that uses waves of energy to sense objects.

revolve To move in a circle on a central axis.

satellite A machine sent into outer space to travel around the planet and obtain information that is sent back to Earth.

tundra A large, flat, treeless Arctic region.

For More Information

Books

Griffin, Annabel. *Weather and Climate*. Truro, England: Hungry Tomato, 2021.

Hicks, Dwayne. *Mapping the Equator*. Buffalo, NY: Gareth Stevens Publishing, 2023.

McCoy, Erin L. *Global Warming*. Buffalo, NY: Cavendish Square, 2019.

Websites

Mapmaker: Latitude and Longitude
education.nationalgeographic.org/resource/mapmaker-latitude-longitude/
Explore an interactive map with global latitude and longitude coordinates.

National Geographic: Altitude
education.nationalgeographic.org/resource/altitude/
Learn more about how altitude impacts every region on Earth.

National Geographic: Atmospheric Pressure
-education.nationalgeographic.org/resource/atmospheric-pressure/
Read more about atmospheric pressure and the way it coincides with altitude.

Publisher's note to educators and parents: Our editors have carefully reviewed these websites to ensure that they are suitable for students. Many websites change frequently, however, and we cannot guarantee that a site's future contents will continue to meet our high standards of quality and educational value. Be advised that students should be closely supervised whenever they access the internet.

Index

A

air pressure, 6, 8–9
altitude, 22, 23
atmosphere, 4, 9, 18

C

clouds, 4, 8, 12, 13
cyclonic climates, 24

E

Earth's axis, 7
equator, 16, 17, 21, 22, 24, 26, 27
evaporation, 10, 18

F

fronts, 14

G

global warming, 29
Gulf Stream, 19

L

latitude, 16, 17

M

meteorologists, 14, 15
monsoons, 20, 21
mountains, 22–23, 26

O

oceans, 10, 11, 18–19, 26

P

permafrost, 26, 28
polar regions, 6, 16, 17, 19, 24, 26, 27, 29
precipitation, 4, 5, 6, 8, 12–13, 20, 23, 24, 25, 26

R

radar, 15

S

satellites, 15, 25
sun, 4, 6, 7, 8, 9, 10, 16, 18

T

temperature, 4, 5, 6, 8, 10, 12, 13, 14, 15, 16, 19, 22, 27
tropics, 6, 16, 19, 20, 22, 24, 25, 26

W

water vapor, 10, 12, 18
wind, 4, 6, 10, 11, 19, 20, 21